NARRATIVE OF THE WRECK AND LOSS
OF THE WHALING BRIG WILLIAM AND JOSEPH

NARRATIVE

OF THE

WRECK AND LOSS

OF THE

WHALING BRIG

WILLIAM AND JOSEPH,

OF MARTHAS' VINEYARD.

BY ELISHA DEXTER, LATE MASTER.

SECOND EDITION.

NARRATIVE

OF THE

WRECK AND LOSS

OF THE

WHALING BRIG

WILLIAM AND JOSEPH

BY

ELISHA DEXTER, LATE MASTER.

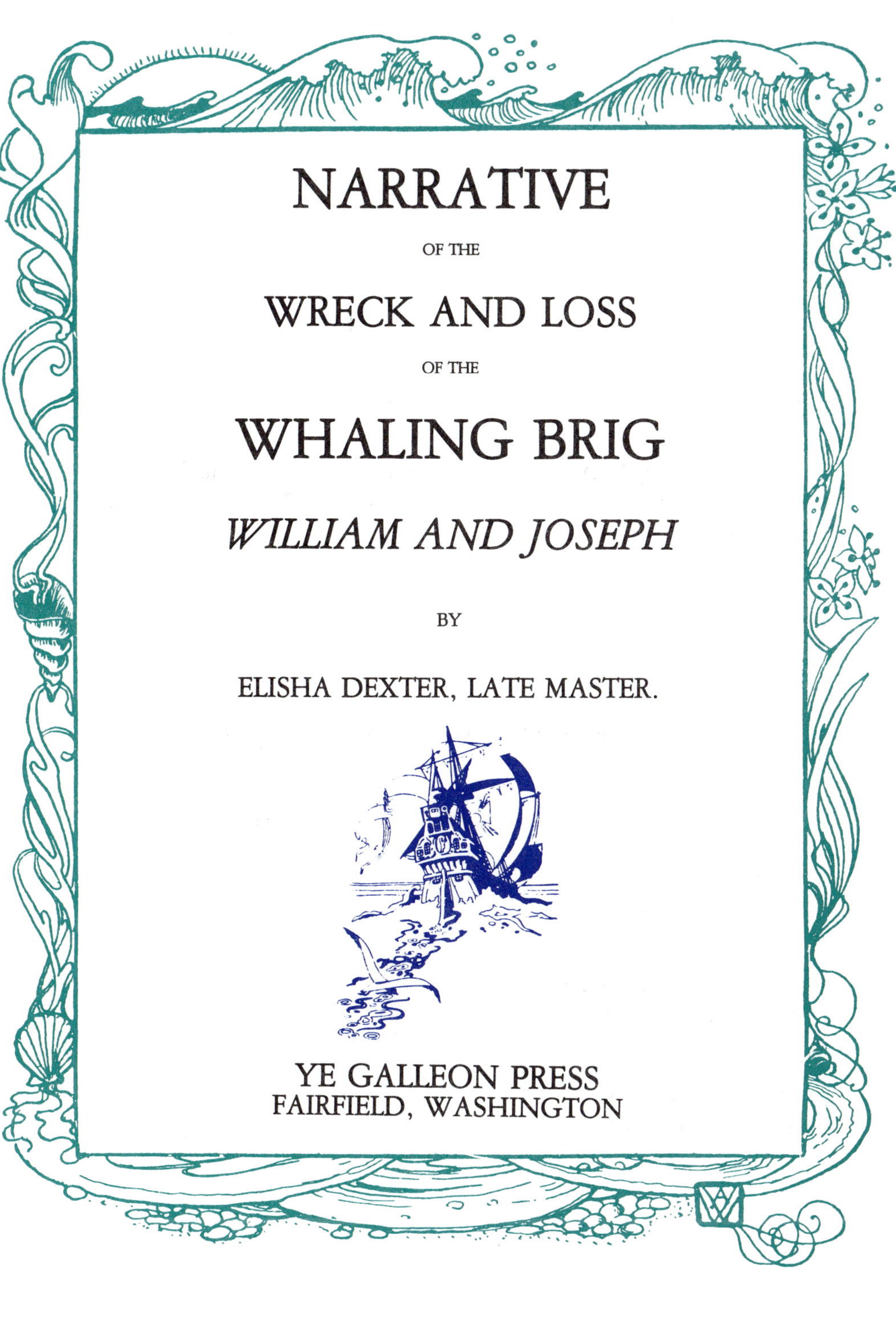

YE GALLEON PRESS
FAIRFIELD, WASHINGTON

Library of Congress Cataloging-in-Publication Data

Dexter, Elisha.
    Narrative of the wreck and loss of the whaling brig William and Joseph.
        p. cm.
        Reprint. Originally published: Boston : C.C. Mead, 1848.
        Includes index.

1. William and Joseph (ship) 2. Whaling. 3. Shipwrecks — North Atlantic Ocean. I. Title.

G530.W7D49   1988   910'.091631   dc 19
ISBN 0-87770-417-1
ISBN 0-87770-418-X

ADVERTISEMENT

As but a small number of the first edition of this Narrative was printed, and its circulation therefore necessarily limited, I have been induced, by some of my friends, to get up this second edition,—the demand being so great,—and in doing which I have spared no pains to have it as accurate as possible. The whole has been rewritten and greatly improved. Some incidents omitted in the first edition, have been inserted in this. Much valuable and additional information has also been incorporated with the Appendix. I flatter myself that the whole is of a character to render it acceptable to the reading public; and especially as it is afforded at the extremely low price of the former edition; and which I hope to offer, by the increased patronage of those who believe that ''truth is stranger than fiction,'' and which fact they will see most fully demonstrated in this Narrative. The Appendix alone is worth the price of the book.

ELISHA DEXTER

Holmes' Hole, April, 1848.

NARRATIVE OF THE LOSS OF

WHALING BRIG

WILLIAM AND JOSEPH,

OF MARTHA'S VINEYARD,

AND THE SUFFERINGS OF HER CREW FOR SEVEN DAYS, A PART OF THE
TIME ON A RAFT IN THE ATLANTIC OCEAN; WITH AN

APPENDIX,

CONTAINING SOME REMARKS ON THE WHALING BUSINESS, AND DESCRIP-
TIONS OF THE MODE OF KILLING AND TAKING CARE OF WHALES.

WITH PLATES

DESCRIPTIVE OF SOME OF THE PRINCIPAL SCENES.

BY ELISHA DEXTER, MASTER.

SECOND EDITION, ENLARGED AND IMPROVED.

BOSTON:

CHARLES C. MEAD, PRINTER, 4 STATE STREET.

1848.

TRANSCRIPT OF SHIPPING PAPER OF

BRIG WILLIAM AND JOSEPH

ELISHA DEXTER, Holmes' Hole, Master.
BENJAMIN MERRY,   do.   Chief Mate.
GERSHOM DUNHAM,   do.   Second Mate, died.
PETER DILLINGHAM,   do.   Boat Steerer.
CHAS. DILLINGHAM,   do.   do.
CHARLES MULLEN, Philadelphia,   do.
FRANCIS COTTLE, Holmes' Hole, Ship Keeper, drowned.
WILLIAM T. WEST,   do.   Foremast Hand.
GEORGE SYLVA, Fayal,   do.
ABRAHAM HUDDESTER, Baltimore,   do.
WILLIAM ARMSTRONG, New York,   do.
JAMES WALKER,   do.   do.
HOSEA GOODSPEED, Hudson, N.Y.   do.   died.
WILLIAM CRAFT, Albany,   do.   do.
JAMES McGUIRE   do.   do.   do.
JOHN GROTTON,   do.   do.   do.
JAMES GOODRICH,   do.   do.   do.
SYLVESTER DAYLEY, Worcester, Mass., Steward, drowned.
JOHN FRANCIS, New Haven, Conn., Cook.

NARRATIVE.

CHAPTER I.

N THE *2d AUGUST, 1840, I PUT TO SEA IN THE* Brig *William and Joseph*, of Holmes' Hole in pursuit of sperm whales. My crew, at that time, consisted of sixteen men, being two full boats' crews, and officers for a third boat. Five more men, to complete the crew of the third boat, were to be procured at the Western Islands. For the first part of the passage the weather was rather bad, with strong gales from S. and S.S.W., bringing us under double-reefed topsails and courses, and causing the brig to leak considerably. But when we had gained the longitude of 50° W. the weather became more pleasant, and the gale abated, though there remained a heavy swell from the S.S.W., caused by the recent high wind. There had been a number of vessels in sight, though we had spoken none. Our course was about S.E. by S., and we stood in that direction till the 25th of the month. On the morning of that day, a little before sunrise, it being entirely calm, I saw upon the tafferel, upon which I had stepped for observation, a small breach in the water, which I judged to be that of sperm whales,—and such they proved. They were going at a quick rate, toward the southeast. I sent off the

larboard and waist boats, (*see Cut*), in pursuit of the fugitives, and about 6½ o'clock they came up with them. We were fortunate enough to obtain three. A light breeze springing up from the S.W., we took the brig down to the whales, got them along side, and by 7 o'clock in the evening we had them safe and snug in the blubber-room. At 5 o'clock the next morning we commenced trying out. They were what we called "school whales,"—very small fish,—the three making only thirty-eight and a half barrels. We continued standing to the E. under easy sail, the weather very fine, and our hopes very high; the eyes of all continually strained in anxious gaze for the prey the whaler covets. On the 28th we saw a number of large whales, but their legs were longer than ours, and they escaped us.

On the 2d of September we ran into Fayal, one of the Western Islands, or Azores, for the required complement of men, and for the purpose of obtaining the usual supply of vegetables, which are there very cheap. These islands, which belong to Portugal, lie between 37° and 39° N. latitude, and 25° and 31° W. longitude. The description of one of them will nearly answer for the whole. The soil, climate, and the character of the inhabitants are much the same in all of them.

They have a very fine climate. From the middle of May, to the middle of September, the weather is very bland and delightful, the prevailing winds being from S.W. to W.N.W.; the first named being the strongest wind, and that which chiefly prevails during this period. The southerly winds, as they take nearly the whole sweep of the Atlantic, bring the most rain; and the easterly winds, as they come from the continent of Europe, bring the finest weather. These winds bring the halcyon days of these sunny isles. The sea is then smooth and delightful. Fayal has the finest harbor of any of these islands. It is safe from all winds except those from N.E. to S.E., which blow directly in,—consequently it is dangerous lying there with heavy winds from those quarters. They are, however, but little visited during the windy months.

The soil of these islands is very fertile, producing abundance of grapes, oranges, vegetables, and Indian corn. They use no manure, except a weed about a foot and a half high, full of leaves and sap, which they turn under the soil, as clover and buckwheat are elsewhere, as a dressing for their crops. Prices are usually as follows: potatoes, 25 cents per bushel; onions, 33 cents per bushel; cabbages, 6 cents per dozen; oranges, one dollar per hundred; eggs, 10 cents per dozen; pumpkins, 25 cents per dozen; and beef, from 5 to 6 cents per lb.

Ships usually procure their supplies through the agency of Mr. Dabney, the American consul, who obtains, what is not to be had at Fayal, from Pico and Flores, two neighboring islands, which are separated from the former by a channel. The quantity taken is usually about 150 bushels of potatoes, and 75 bushels of onions, and other articles in proportion, for each ship. The oil taken by outward-bound ships, is usually left at this place, in charge of Mr. Dabney. The fleet of 1839 landed nearly 6000 barrels of sperm oil. The oil is landed from the ships in large boats; a task which is performed with great ease.

The scenery of these islands is also very beautiful. As you sail by them, at the distance of two or three miles, in the season when the orange groves are in full bloom, and the fields green and verdant, nothing can be more picturesque and delightful than their appearance. The contrast exhibited by the lofty mountains and the deep valleys, is grand and striking. They are, indeed, a perfect paradise; and are extensively known as one of the best residences for invalids on the globe.

Among the most interesting and delightful objects to be seen here, are the gardens of Mr. Dabney, of which he has two—that adjoining his house being chiefly ornamental, and the other, which is a mile distant, being devoted to the raising of produce. The ornamental garden has many fruit trees, and a great variety and abundance of all kinds of flowers, which will grow in that climate, which, from its temperature, must be nearly those of all climates on the globe. It is laid out with beautifully gravelled walks, trimmed with box, and is a most splendid garden. The other contains fifty acres, and has a fine observatory, from which there is an extensive view of the surrounding country, together with the sea-board. In this garden are raised all kinds of vegetables and pines, together with the earlier and better kinds of oranges. The wall which surrounds it, is, perhaps, the most curious thing about it. As there is no stone on the surface of the ground larger than a pea, he has gone four feet below the soil to obtain it. He has there found an abundance of porous stone, with which he has constructed it. As labor is very cheap, (only a shilling a day,) it has cost him but little.

As to the inhabitants of these islands, I speak but the common opinion of all who have ever visited them, when I say that they are a very kind and generous people. Any person may enter their houses, without any invitation whatever, and receive an apparent, and, I doubt not, an actual welcome. Sometimes, with a graceful motion of the hand, in the skilful use of which they resemble the Asiatic races, they will say, ''enter.'' When you have entered, they will, sometimes present you with a ''Jesu Christo,'' which is an image of our Saviour, made of dough. In presenting this, they will kiss it, and you are expected to receive it with great veneration. It is indeed a pity that so gentle and hospitable a people should not possess a purer and a better faith. I speak now of the lower classes, not having access to the higher. They are very partial to the Americans. The word ''Americano,'' sounds well in their ears.

On the 3d, we came out of Fayal, and stood to the W.S.W. of

the islands. We now saw breaches very frequently, which, from the ground on which we were then cruising, and from the characteristic marks, we knew must have been those of sperm whale. But we were not able to make any more than breaches of them. About the 7th we spoke a ship from Batavia, bound for the Mediterranean, out ninety-five days, and supplied with a small quantity of the produce of the islands.

Having spoken several times of "breeches," I propose to show my readers what they are. Many already know; but there is another and larger portion of the public who do not; and for their instruction I give the following description: The whale jumps out of the water, and when he falls into it again, the water flies into the air in the shape of an inverted cone, or sugarloaf. The height of the volume of water thus thrown up, of course will depend on the size of the whale, or rather on his weight and agility. Some, we presume, are better "jumpers" than others. I have seen a column of water, forty feet in height, thrown up by a whale. Now, when it is remembered, that we constantly keep a man on the main-top-gallant-head, to look out, a height which will range from seventy to a hundred feet, according to the size of the vessel; it will be seen that we can discover the object of our search at a very great distance, and can judge pretty accurately the height of the white spray, as it is thrown up against the field of boundless blue in the distance. [*See Cut opposite.*]

Seeing nothing off the Western Islands, we soon after left that ground, and stood into the latitude of 36° N. and longitude of about 24° W. We there saw three sperm whales; but, as the whalemen say, we had *hard luck.* I will here observe, that nine-tenths of the time, this "hard luck" is nothing more nor less than bad management. The excuses are endless. They will say "the whale did not lay right," or "she rolled from them," or "turned away a little too soon," or lay "a little too long." But a good whaleman may generally be known by his having few excuses. If you ask him to account for his failure, at any particular time, he will probably confess that it was his own fault.

We cruised in the above-mentioned latitude and longitude for some time, but saw nothing. On board of my vessel all things went on well. Notwithstanding "hard luck," which is very apt to sour the temper of men, and make a ship a nursery of crabs, all was peace and harmony and good nature. About the time, however, that we were near the line, one of the men, whom we obtained at the islands, was heard to say, that if I did not use them well they intended to throw me overboard. But this was probably a joke. I always use my men well; though I make them do their duty, and know their place.

In a whale ship "patience and perseverance" are the making of a voyage. Without them, the prospect for a full ship is but a poor one. And further,—be sure to give your men plenty to eat. men cannot work anywhere without food, and least of all, will they do it in a whale-ship. They must be kept in good humor or they will not see whales. At the longest, our years are but few, and the voyage of life is soon up; and why should men be debarred of their food, which is the most they get for their labor?

As the season, in this latitude, was getting late for whaling, and there being a fine wind from the northeast, I thought it best to shift my ground and to run down to the Cape de Verd Islands, and to take a look there on my way further South. We made the Isle of Sal, one of the group, on the fifth of October, and cruised off and on for some days, but saw nothing save two schools of black fish, for which, on account of the rugged weather, we did not lower.

The Cape de Verd Islands are but little more than barren rocks, lying about three hundred miles west of Cape Verd, on the African coast. Like the Western Islands, they belong to Portugal. Those of Sal, Mayo, and Bonavista, are famous for the vast quantities of salt which are exported from them. It is made, not by boiling salt water in bottles, or by its evaporation in wooden vats, as in some parts of our own country, but in large shallow ponds on the shore, which are filled with the sea-water by the high tides; and from which, when the salt is made by the solar heat, it is raked into heaps and removed for exportation. Thus large quantities are made with but little labor; and hence the extremely low price at which it is sold.

St. Jago is the most fruitful of these islands; but even that is very barren. They produce some dry grass,—enough to keep life in the donkeys and goats, but not flesh upon them. But little can be obtained there that men will eat, save when they are almost in a state of starvation. The months of July, August and September constitute the rainy season. The great rains then revive the herbage for a short time; but when it is past, the intense heat which prevails for the remainder of the year, withers it all, and both man and beast are famished. Nothing grows there, as food for man, except a little corn and a few beans. I chose the harbor of St. Vincent in which to anchor and recruit, as it is one of the finest in the world, entirely land-locked, and capable of accommodating five hundred sail at a time, in perfect security. I here obtained eight lean goats, for which I gave a barrel of flour.

About the 13th of October, I again got underweigh, and put to sea, with a good breeze from the N.E. We ran under the lee of the Island, and being partially becalmed, as is usually the case near high lands, took a look at Bravo, one of the southwest of the group. Here we fell in with a Sag-Harbor ship, thirty days out, with ninety barrels of sperm oil, and bound to the Islands for hogs. From Bravo we ran out to the latitude of 7° N. and longitude of 27° W.

On the 28th of October, we saw a school of small whales, of which we took three, and which made us about twenty-five barrels. I cruised thereabouts for two weeks, but saw no more at that time.

It now commenced raining more or less every day, and, finally, so hard and with so little intermission, that I began to think the sails would rot upon the yards. The season of the year threatening a continuance of rainy weather, I hauled out as far as 12½ ° N., and found a clearer sky, and a partial cessation of the down-pouring torrent. It was my intention to have staid (*sic*) in this latitude till the middle of December; but returned, however, to my old cruising ground, on the first of the month. We descried a lone whale the third day after our return,—saw him spout twice, "turn flukes," and then disappear forever.

We continued cruising between the latitude of 5 ° and 7½ ° N. from the first of December to the first of February, 1841. During this time we saw whales thirteen times, and obtained 110 barrels of oil; a part of the time the weather being too bad to lower the boats. On the first day of February, I put away for the West India islands, in search of whales and better weather. We had fourteen days passage to Barbadoes. Passing that island, we ran down past St. Lucia to the island of Dominico. Wanting wood and water, I put into Prince Rupert's Bay, which lies on the west side of the north end of the island, to obtain them. We procured them, as also some yams, oranges and sweet potatoes. Very soon there came an order for us to leave, as it was not a port of entry, and they were fearful that we would smuggle with the blacks. So we weighed our anchor and ran down to St. Thomas, in which harbor we anchored on the 1st of March. Here we laid six days, to give our men a chance to stretch their legs.

On the 7th of March we again put to sea, running to the north of the islands, so as to take a look off the Bahamas. On our arrival there, which was about the middle of the month, the ground looking very barren and unpromising, I became very much dissatisfied with the prospect, and put away for the Western Islands, shortening sail every night. I had a very rugged passage till

I arrived in the latitude of 33° N. and longitude of 47° W., when I had better weather. Here we obtained a whale, which made eleven barrels of oil. I cruised in this latitude awhile, but saw nothing more.

On the first of May I put into Fayal to recruit, which, with the passage, occupied about two weeks. In the latitude of 37° N. and longitude of 31° W., I saw a large whale, but did not get him. Two or three schools of black-fish gave us some sport, together with six barrels of oil. Off Flores, a few days after, I saw a large fleet of whalemen, one of which I spoke. He was ten months out, with *five barrels of sperm oil.* I cruised off and on as far as 42½° N. and then from 30° to 42° W. longitude. Here we saw a school of whales, and struck two of them, but the irons drew. The next day we struck a large whale, but he went off with our line, and escaped, after a chase of sixty miles.

I cruised about in this sea, shifting my ground as often as any good seemed likely to come of it, until the 23d of September. During this time I saw whales five or six times, and was at Fayal two or three times for necessities. About the middle of September we obtained a whale, which made us forty-five barrels. Were it not that I decried, in the early part of this narrative, the practice of pleading "bad luck," to cover mismanagement, I would plead it now, for certainly ours was bad luck. At this time we struck a large whale, and again parted the line. Two days after, we were again alongside of another, but the irons bent against his side without entering the flesh.

CHAPTER II.

N THE 23rd OF SEPTEMBER WE PUT AWAY FOR home, with 200 barrels of sperm, and 12 barrels of black-fish oil; low in spirits, for this was a very inadequate recompense for fourteen months of hard toil and incessant watching. Hard fortune pursued us in another form,—the winds became adverse, when we expected the trades. For a season the winds blew from all points of the compass; but presently they became more favorable, and on the 16th of October they were fair, and, as our provisions were low, we crowded all sail to get home as soon as possible. Our sails were now very much worn, and had become so poor that they were hardly fit to set. However, we kept her going, the wind being southeast.

Gradually the wind kept hauling, as well as steadily increasing, until it became a strong breeze at the south. The morning of the 20th opened upon us with the wind, by compass, S. by W., and blowing fresh, though not so fresh as to prevent us from carrying all sail till sunset. The sky, which was cloudless in the morning, began to be overcast about noon. The wind kept steadily increasing, and the flaws became stronger and more frequent. A dark cloud seemed gathering in the northwest. Still we had no suspicion, from any sign in the weather, that a terrible disaster was at hand. Surely ''we know not what day may bring forth.''

About sunset, my ship-keeper, Mr. Francis Cottle, came to me and observed that we should have a fine night, as the sun was shining through the clouds. I thought the same myself. However, as it looked wild and lowry, between sunset and dark, I ordered sail to be taken in, bringing her from all sail set to a two-reefed main-topsail, a single-reefed fore-topsail, whole foresail and fore-topmast staysail. Under this sail we ran before the wind, making good weather of it. About 12 o'clock at night, my second mate, Mr. Dunham, came below, and reported that the flaws of wind came stronger, and gave his opinion that the fore-topsail should be taken in. I had been trying to sleep, but without success, for I felt an indefinable sense of approaching danger,—a dread of something, I know not exactly what. So, being fully awake, and entirely prepared, I very quickly hurried on deck, and told the second mate that we would take in the fore-topsail between the flaws, which we succeeded in doing. The flaws now came increasingly fast; but still, the brig being a good sea-boat, was very easy, and I saw nothing to especially alarm us. There was quite a long sea on, and once in a while it would break—the crest of the sea running into foam. If I remember rightly, I staid (*sic*) on deck till the larboard, or chief mate's watch was called, which was 3 o'clock. I gave the chief mate directions to call me, if it should blow harder, and then went below.

At 4 o'clock in the morning, of the 21st, the chief mate came below, and said that the gale still increased, I directed him to call all hands, and immediately hurried on deck. All hands came up. It was now my aim to get an opportunity, between the flaws, to take in the main-topsail, which we were enabled to do, leaving her to run under the foresail. Suddenly the gale increased, so fast, that I directed the foresail to be hauled up and furled as soon as possible. As they let go the tack of the sail, both of the clew-garnets, being old and rotten, parted, the weather one first, and then the lee one. At the same moment there came a bad flaw of wind, and the foresail went into ribbons in the twinkling on an eye. Our fore-top-mast-staysail was also blown away at the same time.

In this dilemma I resolved to put on the trysail, to keep her to the wind. The wind had somewhat moderated, and had also hauled two or three points, so as to be S.W. by S. As we were hoisting up the sail, and had gotten it about half way up, a squall from W.S.W. struck us, and blew tremendously hard. It seemed as though the whole air had burst into one wild, roaring wind. In the whole course of twenty-six years service on the ocean, I never before saw the like. This squall immediately knocked the brig about one third over. She would easily have recovered herself from this, and righted; but another, and another, and another squall, each fiercer than that preceding it, came in a moment of time, and over she went, keel out, and masts under. [*see cut on page 24.*]

Such a scene as then presented itself, I never before witnessed! It was as dark as night could be,—pitchy darkness,—relieved only by occasional sheets of white foam, while the wind moaned and howled terribly, and the sea ran mountains high. "Oh my God! my God!" was now heard from every part of the wreck; and not the faintest cries for help from those who had cursed and derided their Maker for the whole voyage. "Give me a rope, or I shall drown," exclaimed one. "Save me! save me!" cried another, in the greatest distress, and in the most piteous tones. But at such a time each man is for himself. He cannot well help his neighbor. The cries for succor were therefore unattended to. Each man was thinking only of himself, and fearful of letting go his hold, lest he should be precipitated into the deep, which was raging around him. The love of life is very strong. "All that a man hath will he give for his life."

It is worthy of remark, as a proof of the existence of a power above us, and of man's need of His help, in extremities, that when in great distress and their utter helplessness is made apparent, men, who, apparently, never reverentially thought of their Maker, will at once exclaim, "Oh my God!" and call lustily for His assistance.

When the brig capsized, I was standing by the cabin gangway, holding on to a rope which the chief mate had made fast to the rail, in the early part of the night, to hold on by, if need be. I was so encumbered by a heavy pea-jacket, that if I had been thrown into the sea I could not have swum, and must have been drowned. However, by the help of the steward, I managed to get it off, so as to feel quite safe from the prospect of being drowned. For some minutes the sea made a clean breach over me. At one moment I was immersed in its depths, and then would emerge, for a second, until I thought that I should be choked with the water. But God took care of me, and of most the rest of us, so that we might believe in His gracious providences.

Mr. Cottle, the ship-keeper, John Mullen, one of the boat-steerers, together with the steward, having, for greater security, crept up to windward, under the lee of the quarter boat, a violent squall again struck us, and immediately took the boat off the cranes, as though she had been a feather, and knocked all three of them overboard. Only one, Mullen, succeeded in regaining the vessel. The two others sunk to rise no more. The unfortunate steward was a colored man; and Mr. Cottle was a native of the Vineyard, where he left a wife and family to mourn his sudden and melancholy death. The two other boats had been carried away when the brig was first knocked down.

What thoughts of country, home, family, and friends; of the events of our past lives, and the dismal prospects of the future, now rushed into our minds. Years were now concentrated in a few moments. Our situation was distressing in the extreme. Our gladsome hopes of home, which we were rapidly approaching, suddenly, almost in the twinkling of an eye, were dashed beneath the sea, and washed away by its raging flood, while some of our helpless and unhappy shipmates were perishing around us, and we could afford them no assistance!

For some minutes I knew not what had become of the crew. The brig lay upon her beam ends about ten minutes, and then righted, full of water, both masts gone by the deck, and bowsprit broken by the night-heads. But for this even we were thankful. When the survivors mustered on the quarter-deck, we found that two only, the ship-keeper and steward, before mentioned, were missing.

It was now broad day-light. We were lying in the trough of the sea, and the waves were continually breaking over us. We were compelled to hold on the larboard rail with all our might, to prevent ourselves from being swept off by the wind and sea, and in doing which I got my hand between the rail and main-yard, which had been lashed down to the rail, and it was so badly bruised that it did not get well until after I arrived at home.

Such a night as the past, I had never witnessed before. The faces of the crew (poor fellows) were filled with despair. They seemed chilled and hopeless, not knowing what to do, and yet anxious to do something. Even the poor dog seemed fully to comprehend our distressing situation, and with a look of despair whined piteously. Here was an unmanageable hulk in the midst of the broad and angry ocean, her hold full of water, and the sea making a clean breach over her. Our minds could settle down upon nothing. With the greatest part of my crew this was their first voyage; and it will be allowed that this was poor encouragement for them to persevere in a sea-faring life. And yet, so prone are men to forget disaster, and most of all, marine disaster, that I will presume to say, that the greater part of them were off again within two months. One, I know, remained at home, after his arrival, but just one week.

About sunrise the wind began to die fast away, and 4 o'clock in the afternoon—what a change! One could scarce believe that the whole had not been a dream. Man is bound to persevere, while life lasts, and to make the best of every thing. So the first thing we did, after the gale abated, was to fish up whatever things we could that were floating in the cabin. We first got the brig's ensign, which served us for a signal of distress, and was a fortunate prize. Next, our chests; but the contents of them were all washed out, except the second mate's, which was fortunately locked. Next, the drawer of our table, with a hatchet in it, which, although unground, was very serviceable to us. We also saved two of our cutting spades, and with them we commenced cutting holes in the deck. This gave us employment, keeping us, by turns, busy and active. We soon hauled out a barrel of flour, but the water, to make it of use, was wanting.

The deck now soon began to rise, by reason of the casks in the hold being pressed against it by the water within; first starting up at the larboard water-way-seam amidships, the seam growing wider and wider every moment. It now occurred to my mind to lash our spars together, to form a raft, upon which to save ourselves when the deck should part from the body of the vessel. We had a spare main-yard, top-mast, and try-sail-mast, and of these we made a triangle, which, when finished, perhaps might be the means of saving us.

We now began to feel very much the want of water. We had nothing to eat except the above-mentioned dry flour, which, without water, could not be converted into food. Hunger had not taken sharply hold of us as yet; but eighteen hours having elapsed since our last meal. In the afternoon we also hauled up an old topsail from the after-hold, with which, the next day, we made an awning to protect ourselves from the night air.

About 12 o'clock we saw a ship dead to windward of us, about half mast up. We kept our eyes constantly fixed upon her, in hopes that she would drift down to us, as she was under short sail, and wearing ship occasionally. By 4 o'clock she was fully in sight; and we discovered that her fore-top-mast had been carried away, probably in the gale of the preceding night, by which we had been wrecked. We anxiously watched her till night closed in, which shut her out from our view before she could see us; so that we saw no more of her, and thus were bitterly disappointed of our hopes of relief.

The first night after we were wrecked was passed in dreary wakefulness. Fearing that the brig would go down in the course of the night, we tried to sleep on the spars which formed the triangle intended for the raft. But

> *Tired nature's sweet restorer, balmy sleep,*
> *Which comes to all, came not to us.*

Sleep fled from our eyes. Our anxious look-out for a vessel to rescue us from our most perilous situation, kept us awake the whole night. We kept watch, three and three, to make sure that no vessel should pass us. But even if we had been disposed to have slept, we should have been prevented by the unceasing noise made by the surging of the casks in the hold, and their continual knocking against each other and against the deck. The bright, full moon infused a little cheerfulness into us, yet it was a truly uncomfortable night, though nothing to those that followed.

The morning of the 22d at last came, but brought with it nothing to cheer us. Nothing was to be seen in the broad expanse of ocean but our poor, helpless selves. The air was now bland and the sea smooth; but hunger and thirst began to make known their imperative wants. The casks began to work out of the fore-hatchway, but there was not one amongst them containing fresh water. Our fresh water having been all vented to keep it sweet, had become so impregnated with the salt water, that it was no better than that alongside. One four-barrelled cask of water, however, was not so salt as that of the ocean,—perhaps in the proportion of one part of fresh to three of salt. But even this was thus rendered unavailable. We now remarked to each other, that if we could get the barrel of molasses, we should fare sumptuously; when behold, to our great joy, out it came. But, alas! we were again doomed to a most bitter disappointment. The molasses had been displaced by the briny element. Some attempted to eat the dry flour but could not. We were now in strength and disposition reduced to mere children; but we thought that if we could get a little fresh water we should again be men. Our misery was, indeed, great upon us. For myself, I did nothing,—was able to do nothing. The thinking fell to my share; but I was as incompetent to perform mental, as manual labor. Those best disposed and most willing to labor were able to do but little, there being no tools with which to work, but the hatchet without an edge, and a small bar of iron about two feet long. We could drive spikes, with which to strengthen our raft, however, with these tools, whenever we could obtain any to drive.

But whatever may be the circumstances in which you are placed, of one thing you may be assured—time will keep moving. But the day passed wearily away. We found a sixty-gallon cask of corn, which was nearly destroyed by the bugs. A small portion of this was eaten; but the want of fresh water to wash it down, finally compelled us to give it up. When hunger and thirst exist together, the former cannot be easily satisfied alone. Thirst is the most imperious in its demands.

As the night came on, we all gathered under the old sail, our hearts saddened by the awfully dreary prospect that was before us. Hunger and thirst now assailed us in their most formidable shapes. But as great as our hunger had become, it was nothing compared to our thirst; the rage of which had become unspeakably intense. We had often before doubted the praises of cold water, as sung by the devotees of temperance,—but never since that dismal night of sorrow. If any others have doubts on this point, we most sincerely hope that they may never be subjected to a similar mode of conviction. Our tongues were as dry as chips. If we had had water, we should, probably, have felt the pangs of hunger more keenly. But all our wishes were now merged in water, water, water! This second night we set the watch as before, and all of us got more or less sleep.

On the morning of the 23d there was a brig in sight, nearly hull up, but she did not see us. We now went to work clearing away the broken masts and spars, saving every thing which could be of any use to us. Three of our men, also, now set about making a small raft out of the fore-top-gallant-mast and pieces of the fore and main-top-masts, spiking some and lashing others. Upon this raft they rigged a small mast and sail, intending to use it to cut off any vessel that might pass. But she must have been a dull vessel which this sailing raft could cut off, for its speed could never have been more than a mile an hour. I did not like the plan, as I thought it could be of no use whatever; and further, I judged that it would be wiser to use all our materials in the construction of one raft, capable of sustaining the whole crew. And yet, I will candidly confess it as my opinion, that this little raft was, under God, the means of saving my own life, as also the lives of all on the large raft.

We spent this day in search of food and water. A cask of bread was found, but the want of water rendered this also useless. I could not swallow a mouthful, but there were some who did; and to them it was some relief. For myself, I thought of nothing but water. The men drank of the mixed water I have spoken of. The cook also mixed some of it with flour, and passed the dough it made around to the men. The only way in which I could eat mine, was to make it into pills, about the size of those we buy of apothecaries, and then swallow them just as we do pills; though even this could not be done until saliva enough had been formed to assist the organs of deglutition. I found, in accordance with the opinion of Dr. Franklin, that wetting myself in salt water somewhat assuaged my thirst.

Another night, cold and cheerless, came; and the morning of the 24th beamed on the same helpless and despairing men. I now saw indications that the brig would soon sink from under us. The remnant of the deck was coming up, and the casks which had supported her were washing out. While the others searched for water, myself, the mate, and two men, commenced working on the raft, covering it with a portion of the deck, which had come up, together with boards, which washed out of the hold. We constructed it with all possible care, putting on new and additional lashings, and spiking all solidly down; so that, in the course of the day, we had constructed a pretty good raft. The last thing I did was to rig up a mast upon which to set our signal of distress. The other gang, which was at work searching for water, sought but in vain. We were now so weak, by reason of continual watching and fasting, that the whole gang performed, in a whole day, less labor than two of us, before the disaster, could have performed in two hours.

As night was now fast coming on, we divided the old sail amongst us for a covering, and lay down utterly prostrated by hunger and exhaustion. The raft was laid upon the quarter deck, with nothing to keep it down, so that if the brig should sink in the course of the night, the raft would safely float. There being a strong breeze from the north, it was too cold for us to sleep much; but we got through the night as well as we could.

An incident occurred during this night, which may, perhaps, somewhat perplex the unbeliever in the prophetic character of dreams. I stop not to explain the philosophy of this fact. I leave it

for those to whom "no mist is impenetrable—no millstone opaque." Being a humble seaman merely, and one who has had to deal with the stern realities of human life, I go not into matters beyond my depth. Mr. Merry, the chief mate, dreamed that a ship passed close by us, and took not the least notice of us. Immediately upon waking he related this dream to us; and sure enough it literally came to pass. Upon the first dawn of day, on the 25th, it being the second mate's watch, he caught sight of an object a little north of the gray light, which appears at that time in the morning. At first he thought it a cloud; but as it came down he was not long in discovering it to be a ship standing to the south. Immediately he sang out "Sail, ho!—sail, ho!" and in a moment all hands were up, with eyes almost starting from their sockets, to catch a glimpse of the object which we hoped would bring us relief from our perilous situation. To attract their observation, and to cause them to notice us, we shouted at the tops of our voices; rung the cabin bell, which we had saved; beat upon the heads of empty casks, &c. The ship was coming down quickly, but the course she was standing would carry her a mile from us. At this moment, two of the hands jumped upon the small raft, and made sail to cut her off. [*see cut, opposite.*] As the wind was light and the sea very smooth, it was not unreasonable to expect that they would either hear or see us. They were so near, that we distinctly saw a man go aloft to loose the main-royal. She could not have been more than a mile distant, and the small raft midway between the ship and wreck. Yet she kept on her course, apparently taking no notice of us, even if she saw us. How she managed not to see us, if that were the case, I cannot tell; as the sea was not only smooth, and the sky clear, but the flag upon the large raft was at an elevation of full thirty feet, and the small raft, as before stated, not more than half a mile from her. Perhaps, nine persons in ten, would say that she did see us. But others might think it impossible that it should be in the nature of man to forsake his fellows in such an awful extremity of distress as this. But yet it has been done. This crime, without name, has been committed; and therefore it might have been in this case. But we would not judge men too hardly, God knoweth.

But when she had passed away, and had clean gone from us, what feelings of unspeakable horror, desolation and lonesomeness came over us! Our hearts would have died within us, and hope would have fully expired, had we not trusted, that He who heareth prayer, and "who holdeth the winds in his fists, and the water in the hollow of his hand," would not wholly forget us in our utter helplessness.

The unfortunate men upon the raft had hard work to get back again to the wreck,—dejected in spirits, and almost broken in heart, to again mingle with their unhappy shipmates, as badly off as themselves,—as the raft, being constructed without any reference to form, was very unmanageable. Besides, the men upon it had had nothing to eat or drink for five days and nights, and their strength was but that of children. Before they got back again, the wind began to breeze, and the clouds came up from the Northwest quite fast, with every indication of a squall. The sea also began to rise in short swells, and as the deck had risen a good deal, the sea now washed under it, ripping it up more and more, and driving out the casks that had not before moved.

The brig now began to settle by the head, and we saw one principal support, and which had been our ocean-home for the fourteen months past, fast sinking under us. It was about the middle of the day when she commenced settling, and by the middle of the afternoon she had settled forward about two feet. Afterwards, as though struggling for her life, like the unfortunate beings whom she bore up, she came up, so that her bows were about level with the water, and remained so for a short time. Presently she began to settle for the last time, and was soon all under water, except a small piece of her stern.

I took with me, on the voyage, as a part of my stores, twenty-five silver dollars. As the brig was going from under us, and I kneeling in prayer to God for protection and succor, it occurred to me, that to preserve this treasure, now of less value to me than a crust of bread or a gill of water, was a heinous sin; upon which I immediately threw it into the deep. How fraught with awfulness was this moment! Some were weeping, and others praying; and no marvel; for death; which had been hovering over us for five days, now seemed just lighting down upon, to seize us for his prey.

From the beginning, I had resolved to take to the larger raft, when she went down: but from a sudden impulse, for which I cannot account, I now jumped upon the smaller one, whither another man had already gone. I lay alongside the wreck until the other raft should be clear of it, which was about 5 o'clock in the afternoon. Here were now twenty men afloat in the midst of the Atlantic Ocean, upon a few spars and planks, and without food or drink, of which to speak, for five days, but very slightly protected from the chilly air of night, from above, or the surging waters beneath. The wind was now N.N.W., and quite a heavy sea running, which were fast drifting us toward the south.

During the succeeding night, the sea often broke over us, making us very wet and uncomfortable, although the water was quite warm. About 8 o'clock there came up a slight shower of rain. We turned our faces to windward, and held out our parched and thirsty tongues, in order to catch a small portion of the precious liquid. I think that I may have obtained six drops; and O, the value of even one drop of water, to a man dying of thirst. Small as was my draught, it proved a very sensible relief to my tongue, which was swollen and cracked, and felt like a piece of wood. The story of the rich man and Lazarus, (Luke xvi: 19-31,) was continually before me, and especially the most earnest entreaty of the former, ''Send

Lazarus, that he may dip the *tip* of his finger in water, and cool my tongue.''

About 9 o'clock, I told Hany, the man upon the raft with me, that I felt disposed to lie down, and try to sleep, notwithstanding the sea was continually washing over it. We each had a rope made fast to our waists, with the other end fast to the raft, so that we might again recover it, in case we should be washed overboard. I determined to hold on as long as possible. We had rigged a small mast on what we called the forward part of the raft, to which our sail was attached; and I had concluded, as soon as I found my strength fast failing, to lash myself solidly to this, and there to die, with the hope that we might thereby be picked up by some passing vessel, and the manner of our awful deaths made known to our friends. Indeed, such now was the dreadful extremity to which we were reduced, that I had ceased to feel either hunger or thirst,—the sure prelude to a rapidly approaching dissolution; and felt willing to die, if our friends could but get the news; the very recollection of which sensations, even now, causes me to weep.

Upon the larger end of our raft we had rigged a little seat, and upon this I sat down to sing that beautiful hymn of Charles Wesley's, commencing—

> *O how happy are they,*
> *Who their Saviour obey,*
> *And have laid up their treasures above!*
> *Tongue can never express,*
> *The sweet comfort and peace,*
> *Of a soul in its earliest love.*

How healthful and enlivening to the soul is sacred psalmody! The singing of this hymn gave me unwonted tranquility and cheerfulness.

Goodspeed complained to a shipmate near him, of a pain in his bowels, and in a few minutes fell from his seat, and after making some ineffectual attempts to rise out of the water, which was continually washing over him, died in about half an hour. He was a young man of but very few words. He seemed to be impressed with the idea that the only hope of succor was by falling in with some land, and would therefore occasionally inquire, if we should not, probably, soon fall in with Bermuda. But, alas, he was never again to see the land. He was from Hudson, N.Y. Before midnight, Mr. Dunham also fell from his seat in like manner; and while the sea washed over him, and he vainly attempted to rise, for a long time, in his delirium, he mournfully exclaimed, ''Oh my mother, my mother!'' until he also ceased to breathe. He was a native of the Vineyard, where he left one child, a son, and was about fifty years of age. None could render these unfortunate men any assistance, as all were so utterly exhausted, as to be scarcely able to support themselves. To add to the horrors of this gloomy night of suffering and death, a shark, the most voracious of all things which inhabit the sea, several times passed over the raft, which was partially submerged, in a probable attempt to seize one of the men. Several attempted to kill him with a harpoon; but as skilful as they were, with this instrument, no one had strength of arm sufficient to cause it to penetrate the flesh.

Let our situation, at this time, be imagined. Twenty men are afloat in the broad Atlantic, upon a few broken pieces of spars and plank, held together by a few spikes, and frail cords, which are continually chafing away. They have had nothing to eat or drink, of which to make account, for six days and nights, and are now fast perishing with hunger and thirst. Two of their companions have already been washed into the sea, and two more, overcome by their incessant and long-continued sufferings, have just died before their eyes. The sea is foaming and tumbling around. Dolphins are leaping out of the water, as if to mock with their gambols, these dying men. Four times have their hopes of succor, by a sail in sight, been raised, and once, to the highest pitch; and four times have they been dashed into the sea, causing their spirits to sink lower than ever. Visions of home and friends occasionally flit across their mind, but are succeeded by feelings of almost blank despair, so that almost the only words they are now heard to utter, are, "O, my God! my tongue! my tongue!" And except relief speedily comes, the ocean must soon swallow them all up, as it already has their fellows, of which they are all conscious,—most fully realizing that "there is but a step between them and death."

The morning of the 27th, opened with fine, warm weather. We were not so entirely without hope, but that we constantly looked out for assistance, or so entirely without energy as to cease to make exertion to prolong life. There were a few fish swimming around us; and as I had a spruce pole about eight feet long, that belonged to our sail, I sharpened one end of it, and cut some notches in it, in order, if it went through their flesh, to hold them; and with this attempted to catch them; but their skins were so tough, that its point constantly turned against their sides, as the irons sometimes do against the sides of a whale. Mixed with the gulf weed, however, were a few small crabs, about the size of a man's finger nail; and these I ate as well as I could.

CHAPTER III.

AN'S EXTREMITY IS GOD'S OPPORTUNITY.''
We were not all doomed to perish. Succor
was now at hand. About 8 o'clock this
morning, Hany sung out, "Sail ho!"
Looking sharply in the direction he indicated
with his finger, an object was indeed to be
seen; but it was at a great distance, and, as we
soon ascertained, was going from us. Down went our heads again;
and our despair grew deeper and more intense. Presently, Hany
repeated his cry of "Sail, ho!" and surely enough, there was
another sail coming right down upon us: but notwithstanding
there was a good breeze, her motion was very slow,—or at least so
appeared to us, with whom every moment was now an age. I
remarked to my shipmate, that I thought that it was some Dutch
ship "bobbing" on the wind, perhaps bound for New Orleans,
and according to the Dutch fashion, taking it "fair and easy."
Suddenly the wind knocked her off about one point, and as she
was about five or six miles from us, this would make quite a
difference in her approach to us. Fearing now that they would not
be able to see us, we made sail to cut her off; and hope infusing
new strength into us, we likewise attempted to assist our motion by

paddling with our hands; but our minds went faster than our crazy craft. We had to steer the raft, in order to keep it before the wind, and in which we took turns,—one steering, while the other paddled. Our intense anxiety, lest we should again be left as we had been before, now amounted to a mania, so that we incessantly scolded each other for not doing better, although each poor fellow did the best he could. We soon made her out to be a ship, under her topsails and courses, standing closely upon the wind. When she got abreast of us, at the distance of about a mile, she set her colors at her mizzen-peak, shook the reef out of her main-topsail, set her main-top-gallant-sail, and stood on. Presently she tacked and lowered her boat. We now saw that she was a whale ship; and soon her boat was coming down to us with the usual swiftness of a whale boat. In a few moments more we had left our frail craft, and were on board the boat, and soon alongside of that beautiful ship,—beautiful, not by reason of her build, finish or movements, but for the welcome succor she brought us. What a wonderful change was then in our feelings, in a few short minutes,—from the very verge of blank despair, to the height of extatic (*sic*) joy! I felt quite smart, and wonderfully resolute all at once; became quite a man again, and refused water till we should be on board.

No sooner were we on board, than they gathered around us to hear our sad story. The ship proved to be the *Triton*, Capt. Bowen, of Warren, R.I., from New Zealand, with 2000 barrels of right whale, 200 barrels of sperm oil, and 22,000 pounds of bone; last from Rio Grande, where she had been to recruit. Capt. Bowen belonged to Fair Haven. So we were almost neighbors.

Capt. Bowen informed me, that he had felt an anxiety that morning, for which he could not account. He walked the deck in great disquietude of spirit, yet not knowing what troubled him; and that when he first descried us, he supposed us to be deserters from some ship. Upon my informing him, that a larger raft, with the remainder of my men, could not be far distant, and requesting him to go in pursuit of them, he immediately consented to do so. The ship was at once hauled on the wind, and men were stationed aloft to look out for them.

Capt. Bowen now advised me to take a little physic, but which I declined, fearing that it would make me weaker. Upon my calling for water, the steward presented me with a pitcher-full, of which I drank about three pints. But such was the extreme weakness of my stomach, that I soon threw it up. At dinner I eat only a small piece of bread, well soaked in water. Some gin and water was also offered me, but its very smell was offensive to me.

Presently we had the glad news below, that the lookout saw the large raft about eight miles to windward; and by 4 o'clock in the afternoon we had them all on board. I told the captain that I would keep out of sight, in order to hear what kind of a story they would tell; and while Mr. Merry was informing him (standing by the cabin gangway) when and where he last saw me, to his utter astonishment I made my appearance from the cabin. The captain immediately allowanced the men for water; and it had been well had he done the same by me; for I injured myself with the too free use of it. Some supper was announced, and the warm tea did much to restore me; while the soft biscuit of the cook's own baking, tasted most delicious. But my throat was so sore, that it was difficult for me to swallow.

Never did a rescue come more opportunely than this. For myself, I could not have lived exceeding twenty-four hours more, at the longest. Two others could not have lived more than six hours; while several could not stand; and one was in a state of delirium. The lashings of the large raft were also fast chafing away, so that it could not have held together more than a day longer.

How great were the contrasts which we had just experienced. I think that our brig's company met with two as great changes in the course of seven days, as men on land or water can,—from a vessel that has completed her cruise, and is within a few days sail of home, to a complete wreck, and dependence for life upon a few frail spars and planks,—and again from these spars and planks, with death just ready to fall upon us, to a good, tight, well-manned ship, with enough to eat and drink, good beds on which to repose our weary and aching limbs, and kind hearts to watch over us. We most heartily acknowledge the hand of God in our signal deliverance, and were exceedingly thankful for the same. I could but continually exclaim, ''O my God!''

My appetite soon returned, and when it did, it was difficult to satisfy it, without injuring myself. During the first two days, my inclination to sleep was so great, that I slept almost incessantly. On the second day after our rescue, our faces and hands also swelled a great deal.

After a passage of six days, during which fine and pleasant weather prevailed, and nothing occurred worthy of special narration we arrived in Warren, on the 3d of November; thankful to God for all His mercies. During the six days we were on board the *Triton*, every necessary attention, due to unfortunate and suffering men, was paid us: for which I would here publicly acknowledge our obligations to Capt. Bowen, and his officers and crew.

At 6 o'clock the next day, I arrived in the cars at New Bedford, and the day following took the steamboat for Holmes' Hole, where I arrived to tell the tale of my late terrible disaster, with my own lips.

I found my wife and family well. But I have here a fresh leaf to add to the story of my loss. I owned one eighth of the brig, and supposed it insured, as it had been the voyage before. But I was mistaken. Not a cent was insured; and my whole interest in the vessel, including my share of the oil, as master, was a total loss.

All that I received for an outlay of $1000, twenty-seven

months before, and of another $1000, fifteen months before, and my time for the whole period, was my share of the first voyage, consisting of ninety-three barrels of oil. My loss, in the twenty-seven months, was about $2300 of hard earnings. From Dr. Yale, Capt. Bartlett Allen, and "a friend," of Holmes' Hole, I received generous aid, and for which I here beg publicly to express my thanks.

I now am penniless. I put every thing I had afloat, hoping to receive it all back, with large profits, for my adventure, besides something considerable for my "lay." But I have lost it all; and with gray hairs, and a shattered constitution, I am now compelled to commence life upon the land, anew. And now, being no longer able to follow the seas, I am trying to turn even my bitter misfortunes to some account, by the sale of this "NARRATIVE."

APPENDIX.

*HAVE BEEN ENGAGED TWENTY-TWO YEARS,* man and boy, in the Whale Fishery. I commenced at the age of nineteen, after having sailed four years in the merchant service. My first voyage was from New Bedford, in a small brig,—a "cranberry-pudding voyage,"—as an Atlantic Ocean voyage is termed by the sailors. These are of shorter duration than other whaling voyages; but their shortness is their chief recommendation; and they are made at comparatively small expense. After making two of these voyages, I went to the False Bank, in the ship *Com. Rogers*, for right whales. We filled the ship, and returned to port, in eight and a half months. Whales were then more plenty, and also more easily taken than they are now. A ship was then filled in less than a year; while it now takes from two to four years for that purpose. It is a late thing even for ships to go further than the "Tristan ground" for right whales. But now they find the best ground on the Northwest coast of South America, and South of New Zealand, and sometimes sail round the world in the performance of a voyage. The longer, or the Pacific voyages, are the most profitable, they being surer than the short, or

Atlantic voyages. But they require a corresponding outlay.

In whale ships of the same size, the number of men varies but little. Six men are always appointed to a boat, one of whom is "the boat steerer." The number of men appointed to keep the ship, while the boats are out, depends on the size of the vessel. Thus, a ship of 300 tons, carrying from 23 to 2700 barrels, requires about five men, which, with three boats' crews, amount to twenty-three men and boys. A larger ship, carrying four boats, requires from thirty to thirty-two men. The largest ships in this business are out of New London, Conn., some of which are from 5 to 600 tons. Twenty-five years ago, few ships carried more than two boats' crews. Whales were then so plenty, and so easily taken, that a ship with but two boats, could go to the Pacific, and return deeply laden with oil, in from eighteen to twenty-two months. It is reported of a very fortunate and enterprising whaling captain of Nantucket, that in sixty months, including the period he lay at home, fitting, he made three voyages to the Pacific, and filled his ship each time,—in all 6000 barrels of sperm oil. Once he lay becalmed thirty days, during which time he took and stowed away 1000 barrels.

There have been most important alterations, of late years, in the mode of fitting out ships. A ship with a crew of twenty-four men, destined for a voyage of three years, will take 200 bbls. of meats; 200 bbls. of flour, baked and unbaked; 1200 galls. of molasses; 1000 lbs. of coffee; 500 lbs. of tea; 500 lbs. of sugar; besides large quantities of rice, peas, beans, cod-fish, and other articles. I may safely say, that it now costs twice as much to fit a ship, as it did twenty-five years ago. Formerly, they also recruited with much greater care, and at less expense than they now do. Terrapins were then abundant at the Gallipagos (*sic*) Islands; and all kinds of fresh provisions were plenty and cheap. Provisions can now be obtained at any place of note, either at the islands or on the main; but not so cheap as formerly. As increased demand has raised the market. As oil was worth quite as much twenty-five years ago, as it now is, it is apparent that the profits of the business are now far less than they were then. Whales are not only less numerous, but also much shyer than formerly, which greatly enhances the expense and danger of their capture. Hence those who pursue them must be more expert and daring, or a voyage will be easily lost. Ships have sometimes been cut off, and parts of their crews murdered, by the natives of the isles of the Pacific, who are called "Kunackers," by the sailors. But this never occurs where there are missionaries stationed. These humble men are a greater protection to our commerce in the Pacific, than our whole navy would be. But we have to pay something for this protection, however, as the natives under their care are more intelligent than others; they know the value of property, and will not sell, but for its actual worth; while articles can sometimes be obtained from those in a savage state, when we dare trade with them, for a few trinkets or other trifles.

The North Atlantic and the Greenland Seas were the principal whaling grounds of both the Americans and Europeans, previous to our revolutionary war. But now whales have almost entirely disappeared from those seas, so that they are wholly forsaken by the Americans, and mostly by others also. The whale fishery of Great Britain is almost ruined. The town of Hull, in Yorkshire, which formerly owned upwards of seventy sail of whalemen,—as many as Nantucket or New London now owns, has now but four left,—while the whole nation have now but forty-four sail engaged in this business. The whale-fishery of Holland, which was once quite considerable, when the fish were taken in the Northern Seas, is now about annihilated. The Dutch now mostly import their oil from the United States.

France and Prussia have a few whalemen; but the business is

now mostly in the hands of that small fraction of the people of the United States, who reside upon that portion of our northern seaboard, which lies between Salem, Mass., and Long Island,—a small part of four States. These have 687 sail engaged therein, mostly ships, manned by nearly 20,000 men. Of these, New Bedford has 254 sail; Fairhaven, 52; Nantucket, 71; New London, 73; the Vineyard, 11; the Cape, 22; Rhode Island, all together, 35, of which Warren alone has 21; Long Island, 77 sail, of which Sag Harbor has 57. The produce of this fishery to Massachusetts alone, last year, amounted to the sum of about $10,000,000. That of other places probably produced in as large a proportion; from which the immense importance of this single branch of business may be seen.

I will now describe the most stirring event in the life of a whaleman,—the chase and capture of his mighty prey,—"the Leviathan" of the deep.

In the first place, while the ship is on whaling ground, men—generally three—are stationed aloft to look out for the whales; one on the fore-top-gallant cross-trees; one on the main-top-gallant cross-trees; and one on the fore-top-gallant yard. One of them is usually supplied with a good spy-glass. All the time they are anxiously scanning the horizon. The man who first discovers the spout, or breach, cries out, from aloft,—

"There she breaches!"

The officer commanding the deck, at the time, inquires in a loud voice, "Where away?"

"Two points off the lee-bow," answers the look-out.

"How far off?" again inquires the officer.

"About six miles, Sir," is the reply.

"Put your helm up," says the officer to the man at the helm.

"Square the yards—well the main-yard—loose the fore-top-gallant sail—keep a sharp look-out aloft, there!''

''Aye, aye, Sir!'' answers the man spoken to.

''Keep her straight,'' now cries the officer to the man at the wheel.

''There she breaches!'' repeats the look-out from the top.

''Where now?''

''Right ahead.''

''Do we near them?''

''Yes, Sir, we near them fast. There she blows! there she blows! sperm whales lying like logs upon the water.''

''Down from aloft there!'' cries the master. ''Haul up the mainsail—haul back the main-topsail—in lines.''

''There she blows,'' again repeats the look-out. ''There is a large whale in the school; I see one that makes a large spout.''

''We'll try to get that fellow,'' says the master.

''All ready; shall I lower, Sir?'' cries the mate.

''Lower away, as quickly as possible,'' says the master; and lower they do.

''Shove astern as quickly as possible, and down to your oars,'' says the officer of the boat. The men spring to their oars, and the boat moves with great velocity.

''Now, Dick, bear ahand with your craft, and down to your oar. Now spring, all of you, spring. There are a hundred barrels of oil in one carcass, yonder. Spring, might and main. A bottle without rum, ten pounds of tobacco, and a cap for aunt Jenny into the bargain,—and if she spouts twice more, she is ours.''

They have now neared the whale, and are at the proper distance.

''Stand up, Dick,'' says the master to the harpooner. Dick obeys the call; and the boat is guided by the skilful steersman up to the whale's hump. Dick now darts his harpoon into the whale—''*to fasten*,'' is the technical term. The barbs of the harpoon are like those of an arrow—having penetrated the flesh, they cannot easily be drawn out. "Stern all," shouts the officer of the boat; which the men do with a most hearty good will, experience having shown them, that the whale will sometimes cut dangerous freaks with her jaw and flukes. The "stern all," which signifies to pull the boat clear of the whale, stern first, is instantly followed by a return to the attack; and now the officer goes into the head of the boat with his lance, and "the boat-steerer," or harpooner, takes his place in the stern. Sometimes a single thrust of the lance will immediately despatch her, so that she will not run a rod. At other times she will run for hours, even for a whole day, dragging the boat after her with immense swiftness. Sometimes the irons draw, or the line parts, or they are compelled, by the approach of night, to sever it, and she is lost. Sometimes, also, she turns upon her pursuers, with great rage, throwing the boat into the air as though it were a feather, or snapping it in pieces with her mighty jaws, as a dog would a chip, and in which the men are sometimes killed. But it is a singular fact, that when a man is in the water a whale will not touch him. Probably he is too small an object to attract her attention. The boat only, as the supposed enemy, is made the object of attack. A man may be swimming in a school of whales, and be thrust against their huge sides, and they will take no notice of him. Finally, if the whale is killed, she is taken alongside, and if she is a very large one, in great triumph.

The sensations of those who approach a whale for the first time, are not very agreeable; and sometimes, was not a "green hand" more afraid of the officer of the boat, than of the whale, nothing could be done; for a whale fifty to a hundred feet long, is a truly formidable looking opponent. It is not easy to laugh at such a foe; and invincible fearlessness, as well as great dexterity, are indispensable to success.

The first time that I was ever alongside of a whale, was in the year 1819. She was of the sperm species, about seventy feet long, and very vicious. She had been playing antics with Capt. Covell, of the brig. *President*, of New Bedford, having just taken out all three of his lines, with the exception of about forty fathoms. There she lay, lashing her tail with rage, opening her monstrous jaws, which were full twenty feet long, and then again snapping them together with a terrific noise. You could scarcely hear your own voice for the confusion. The two captains now agreed to partnerships, so far as this whale went; and we approached for the fearful attack. In the boats, it was "pull ahead—pull ahead! stern all—stern all! lay this way—lay that way—*why* don't you lay me on!" Finally, the deadly strife terminated by killing her, and taking possession.

The killing of a right whale does not differ materially from the killing of a sperm whale; but the former has some habits which the latter has not. The sperm whale never makes a noise. The right whale is very noisy, sometimes bellowing like a bull, when you get fast to them. During the night, when it is calm, they will sometimes make a noise, resembling the ringing of a brass kettle. The right whale is also a more nimble fish than the sperm. The mouth of the right whale is furnished with the "whalebone;" while the jaws of the sperm whale are furnished with immense teeth. The sperm whale is sometimes destructive even to ships. The ships *Essex,* of Nantucket, and the *Two Generals,* of Charleston, S.C., were both stove by them. There is also a very large species of whale called the "fin-back;" but which is rarely if ever taken, as they move with such immense velocity, that no boat could approach them if they were aware of it; and if struck by surprise, a boat could not hold on to them for a moment.

I will now describe the manner in which a sperm whale is disposed of, after she is taken alongside.

The first things to be done, are to furl the mainsail, and to hoist up the boats on the cranes, and then to set the winding tackles. As the purchase is a strong one, the blocks, which are called "cutting blocks," are very large, being from fifteen to eighteen inches in size; while the "cutting falls," or ropes reeved through them, are about four inches. The lower blocks have straps about four feet long, and on each of these straps there is a large hook weighing from eighty to one hundred pounds, called "blubber hooks." The winding tackles being ready, the captain, with a spade, cuts a hole in the blubber, to receive the hook, and then calls out to the men, who are ready to heave at the windlass, "Hoist away, there! bring to, and heave!—so! avast heaving there!" During this time, all are busy at work; some cutting off the "junk;" others heaving at the windlass, and others "breaking out" the blubber-room, in order to fill it with blubber. The "junk" is cut off separately, and is hoisted in whole, [*see cut, page 54.*] These pieces of junk, just cut off, are called "blanket pieces." If the whale is not large, the "case," or head is hoisted in entire. The contents of the case is called "head matter," and is the most valuable part of the fish. some heads yield over 20 barrels, or 600 gallons of this rich matter, and which may be worth $500. In appearance and color it resembles the core of a ripe water-melon.

The blubber being now stowed away in the blubber-room, or on deck, the "trying out" commences. Two large kettles, containing from 80 to 140 gallons each, set in a furnace on deck, are used for this purpose. Alongside the furnace is placed a large copper receiver, called "a cooler," into which the oil is dipped from the kettles, in order to cool, and from which it is removed to the casks in the hold, by means of a leather hose. This process is continued night and day, till it is finished; during which the ship is filled with smoke, and the hands covered with grease and dirt; and if the highly improper practice is allowed on board the vessel, they

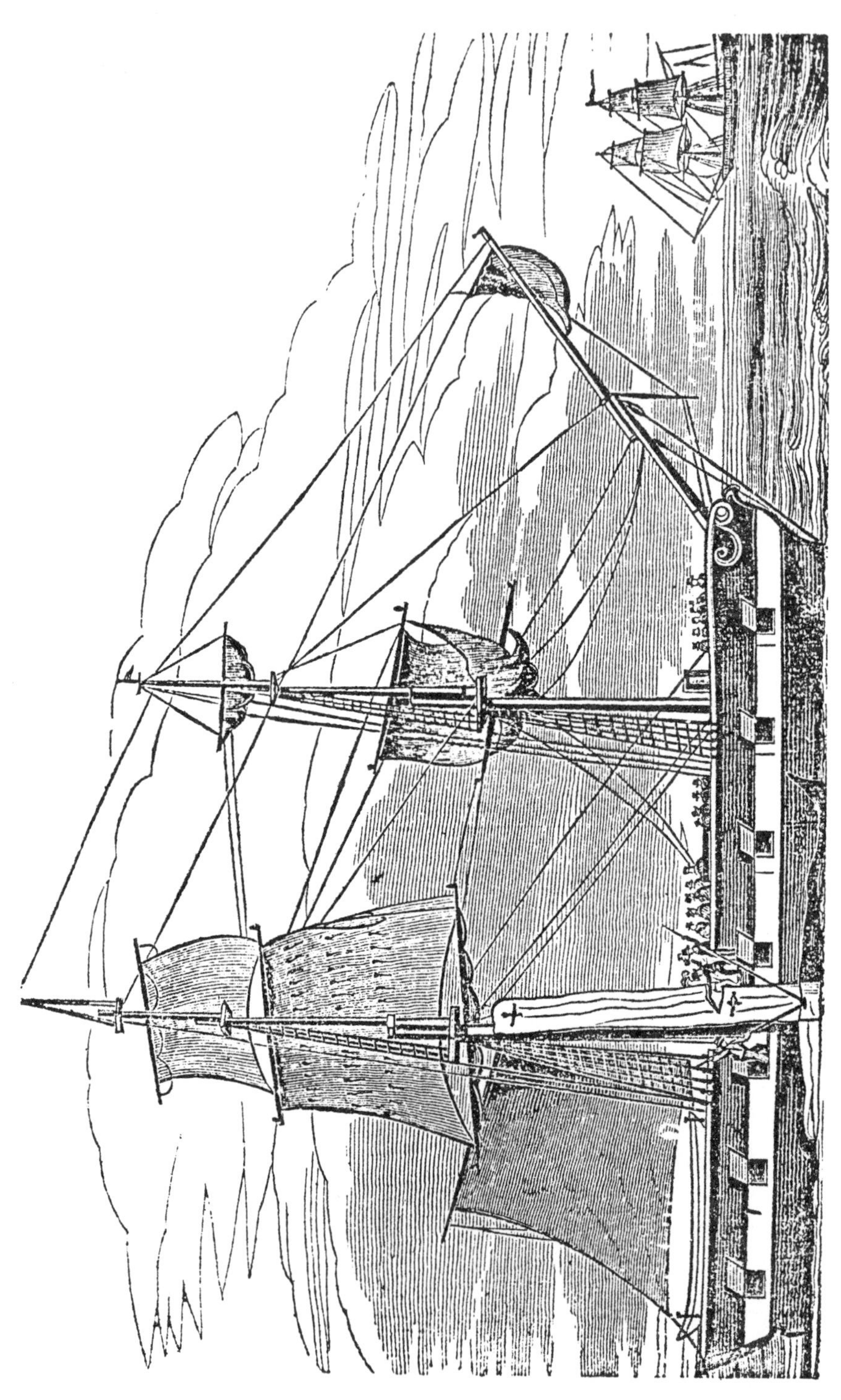

are often in very heavy arrearages for sundry oaths and imprecations. At night the flames light up the ship, making the darkness visible, and night hideous. While the oil is trying out, as it is now as sweet as lard, that important official, the cook, usually avails himself of the opportunity to fry a large quantity of "dough nuts," with which to regale the crew after their severe trial.

While the "trying out" is going on, two hands are employed at the cutting tubs, in mincing. The knife used for this purpose is about two feet long. The blubber is laid on what is called a "mincing horse," and is cut into pieces about eighteen inches long and six inches wide, which are called "horse pieces." Much care must be used in this process. The oil may not be sufficiently tried, or it may be tried too much; as is the case with hog's lard. Much care must also be used to see that the casks in which the oil is stowed, are in a proper state. Oil is sometimes essentially injured in quality, through carelessness in the preparation of the casks. The sediment deposited in water casks, if not cleansed out before the oil is put in, will injure it to the amount of five cents per gallon.

The right whale is much larger than the sperm. The largest sperm whale ever known to have been taken, made a hundred and thirty barrels of oil. The right whale, however, has been known to afford as high as two hundred and fifty barrels. Four of such fish would fill a small brig.

''A system of signals,'' necessarily obtains on board of whale ships. I here subjoin that which I adopted for my own use. —

Whales ahead. Down jib.
  " *astern.* Haul up the spanker.
  " *between the ship and boats.* Flag half-mast.
  " *on the weather bow.* Haul up the weather-clue of the foresail.
  " *on the lee bow.* Lee clew of the foresail.
More whales and a better chance. Flags on the fore-top-gallant mast-head, and peak of the spanker.
Whales on the weather-beam. Mizen topsail aback.
  " *on the lee beam.* Keep the ship off, and luff her up again.
  " *too near to keep off.* Signal to come on board. This signal is made by standing on the top-gallant yards, and holding the flags in your hands.

COLOPHON

The Elisha Dexter NARRATIVE OF THE WRECK AND LOSS OF THE WHALING BRIG
WILLIAM AND JOSEPH was printed in the workshop of Glen Adams, which is located
in the sleepy country village of Fairfield, Washington, southeast Spokane County in
Washington state and one township removed from the Idaho line. The 1854 text
was set by Dale La Tendresse, using a model 7300 Editwriter computer photosetter.
The type face is Garamond size 14 on 17 with running heads and page numbers in
14 point Yorkshire. Camera-darkroom work was by Sylvia Fenich using a 20x24 inch
600C DS (Japanese) camera and a 25-inch LogE automatic film processing machine.
The film was stripped into flats by Sylvia Fenich. Indexing was by Glen Adams.
Paper stock is 70 pound Island Offset. The printing plates were made by David
Hooper, who also printed the sheets using a 28-inch model KORS Heidelberg press.
The sheets were folded by Garry Adams using a 22x28 three-stage Baum air-
operated folding machine. Assembly and binding in paper covers was by Jessica La
Tendresse. This was a fun project. We had no special difficulty with the work.